Early Elementary Science Education

Dr. Wade

Table of Contents

Summary

Primary education is an important part of a child's education. It provides a solid foundation of science knowledge and skills that children can build on in later years. The goal of elementary science education is to introduce children to the basic scientific concepts, tools, and methods used to study the natural world. Through this education, children learn about the world around them, how it works, and how science can be used to solve problems and improve life. And scientific instruments. Children learn to observe and draw conclusions based on evidence. You will also learn to use tools like magnifying glasses and thermometers to collect data.

Other important subjects in early elementary education are forces and motion, magnetism and electricity. Children learn about different types of forces, such as gravity and friction, and how they act on objects. You will also learn about the properties of magnets and how they can be used to generate electricity.

Science literacy also includes topics such as plants and animals, habitats and ecosystems, food chains and webs. Children learn about the life cycles of different animals and plants and how they interact with their environment.

In addition, early childhood education includes subjects such as the human body, the senses, time and climate, earth and space. Children discover the different parts of the body and how they work together and how we perceive the world through our senses.

In early elementary education, children learn about the construction process and concepts of simple machines, structures, and materials. You will learn to apply scientific concepts to real-world problems and design effective and efficient solutions.

Hands-on activities and experiments are an important part of science education in elementary school. These activities allow children to apply scientific concepts and methods to real-world problems and develop critical thinking and problem-solving skills. Children are also encouraged to participate in science fairs and other competitions that will help them develop their communication and presentation skills.

First grade science education offers children many benefits beyond academic knowledge. Children also develop important social and emotional skills through science classes. Participating in hands-on activities and experiences helps children build confidence and self-esteem.

In summary, early primary education is essential for child development. It provides a foundation of science knowledge and skills that children can build on in later years. By introducing children to the basic concepts, tools, and methods of science, elementary science classes help children develop their curiosity, creativity, and critical thinking. Through hands-on activities and experiments, children develop problem-solving skills and learn to apply scientific concepts to real-world problems.

Preface

Science education is an essential part of children's education. It provides a foundation of knowledge and skills that children can build on in later years. Primary education is especially important because it lays the foundation for a child to understand the world around them. In this foreword, we will discuss the importance of elementary science education and its benefits for child development.

Early primary education gives children the opportunity to discover and learn about nature. Helps children to develop their curiosity, creativity and critical thinking. By introducing children to the basic concepts, tools, and methods of science, early elementary education helps them understand how the world works and how science can be used to solve problems and improve the quality of life.

Science education is also important because it is relevant to everyday life. Children come into contact with science in everyday life, be it checking the weather, cooking or playing with magnets. Early primary education helps children make sense of the world around them and gives them a framework to understand the complex systems that make up our planet.

First grade science education offers children many benefits beyond academic knowledge. It helps children develop their social and emotional skills, including confidence, self-esteem, teamwork and communication. Through hands-on activities and experiments, children learn to collaborate with others and develop their problem-solving skills.

In addition, early primary education promotes diversity and inclusion. By learning about the different types of animals, plants and ecosystems, children develop an understanding of the diversity of life on our planet.

Additionally, early primary education can inspire children to pursue careers in science and technology. By introducing children to the wonders of science from an early age, we can nurture their interest and passion for the subject.

Despite the importance of science literacy in early childhood, science education for young children presents many challenges. One of the biggest challenges is finding ways to make science lessons fun and interesting for young children.

Another challenge is the lack of resources available for science literacy. Many schools lack the means or resources to provide quality science education to their students. This can make it difficult for teachers to design engaging and effective science lessons.

Early primary education is essential for child development. It provides a foundation of science knowledge and skills that children can build on in later years. By introducing children to the basic concepts, tools, and methods of science, elementary science classes help children develop their curiosity, creativity, and critical thinking.

While teaching science is challenging for young children, the benefits of early elementary education are clear. We hope this book provides teachers and educators with the tools and resources they need to create engaging and effective science lessons for their students.

Prologue

Welcome to the world of science literacy! This book is designed to give teachers and educators the tools and resources they need to effectively teach science to young children. In this prologue, we introduce you to the world of science literacy, including its challenges and opportunities.

Primary education includes science classes for children from kindergarten through third grade. At this stage, children develop their understanding of the world around them. They are curious and eager to learn more about nature, and science offers them ways to experience and understand the world.

Primary school science education covers a wide range of subjects, including life sciences, natural sciences, and earth and space sciences. Children get to know animals, plants, rocks, weather and other natural phenomena. You will also learn the scientific method, including making observations, asking questions, formulating hypotheses, and conducting experiments.

Teaching young children science can be difficult for many reasons.one of the biggest challenges is finding ways to make science lessons fun and interesting for young children. Children this age have short attention spans and may not be interested in sitting still for long periods of time.

Another challenge is the lack of resources available for science literacy. Many schools lack the means or resources to provide quality science education to their students. This can make it difficult for teachers to design engaging and effective science lessons.

Additionally, some teachers may feel intimidated by the learning and may not have the confidence and knowledge to teach the material effectively. This can lead to a lack of enthusiasm for science in young children and a missed opportunity to spark interest in stem subjects.

Despite the challenges of teaching young children science, there are many ways to make science education interesting and effective. One of the best ways to engage children in learning is through hands-on activities and experiments. Children learn best when they are actively involved in the learning process, and hands-on activities give them the opportunity to explore and experiment with scientific concepts.

Another option is to integrate science with other subjects such as math, literacy and social studies. By combining science with other subjects, teachers can help children see the connections between different subjects and better understand how they all work together.

Finally, early primary education offers opportunities to promote diversity and inclusion. By learning about the different types of animals, plants and ecosystems, children develop an understanding of the diversity of life on our planet.

Early primary education is essential for child development. It provides a foundation of science knowledge and skills that children can build on in later years. By introducing children to the basic concepts, tools, and methods of science, elementary science classes help children develop their curiosity, creativity, and critical thinking.

Forward

Science education is crucial for our future and starts in the first grades. Science education in elementary school is particularly important as it provides children with a foundation for further learning and careers. This book is a great resource for teachers and educators who want to transform the lives of children by providing them with engaging and effective science lessons.

As a scientist and educator, i am pleased that in recent years more emphasis has been placed on scientific education. The importance of science education cannot be overstated. We live in a rapidly changing world that forces us to constantly adapt and innovate.

Early primary education is particularly important because it helps children develop curiosity and an appreciation for nature. By introducing children to the basic concepts and methods of science, we can help them develop critical thinking and problem-solving skills that will serve them throughout their lives. Additionally, early science education has been associated with better academic performance and better long-term outcomes in stem subjects.

This book contains a wealth of information and resources for teachers and educators who want to provide effective and engaging science instruction to young children. The chapters cover a wide range of topics, including life sciences, natural sciences, and earth and space sciences. Each chapter provides background information, ideas for activities, and resources for further exploration. The authors did a great job balancing the need for scientific accuracy with the need for age-appropriate and engaging content.

One of the things i like about this book is the emphasis on hands-on learning. Children learn best when they are actively involved in the learning process, and hands-on activities give them a great opportunity to explore and experiment with scientific concepts. The authors have included many ideas for activities that are easy to do and require only simple materials.

Another strength of this book is its emphasis on diversity and inclusion. Nature education offers children the opportunity to learn about the diversity of life on our planet and to promote respect and understanding for all. The authors have included information and activities that highlight the importance of diversity and the contributions of people from different backgrounds to the scientific community. In conclusion, i highly recommend this book to any teacher or educator who wants to make a difference in children's lives by providing them with engaging and effective science education. The authors have done a great job providing information and actionable resources based on scientific accuracy and age-appropriate content.

Science education is fundamental to our future and starts with the youngest members of our society. By providing young children with effective and engaging science education, we can help ensure a bright and prosperous future for future generations. I hope this book will inspire teachers and educators to take up the challenge of providing children with quality science education and transforming their lives.

Chapter No. 1

What Is Science?

Science is a systematic and experimental approach to gaining knowledge about the natural world through observation and experimentation. It's a way of understanding the world around us through the use of rigorous methods and logical reasoning. The purpose of science is to explain natural phenomena, predict future events, and better understand the universe.

Science is a broad field that encompasses many disciplines, including biology, chemistry, physics, earth sciences, and astronomy. Each of these disciplines has its own methods and areas of study, but they all share a commitment to rigorous observation, experimentation, and evidence-based thinking.

One of the main characteristics of science is that it is based on empirical evidence. Scientists use observations and measurements to collect data about the natural world and then use that data to develop and test hypotheses. Hypotheses are tentative explanations of natural phenomena and are tested through rigorous experimentation and observation. If a hypothesis is supported by evidence, it can be accepted as a scientific theory.

Scientific theories are established explanations of natural phenomena that have been thoroughly tested and supported by evidence. However, theories are always subject to revision or rejection as new evidence emerges to challenge them. This flexibility is one of the strengths of science, as it allows us to refine and improve our understanding of the natural world.

Science is based on logical thinking in addition to empirical evidence. Scientists use deductive and inductive reasoning to make hypotheses and predictions about the natural world. Deductive reasoning is the use of general principles to make specific predictions, while inductive reasoning is the use of specific observations to develop general principles. Both types of reasoning are important in science, and scientists must be careful to use them correctly.

Another important characteristic of science is its openness and transparency. Scientists are expected to present their methods and results honestly and accurately, and to share their findings with the scientific community. This allows for peer reviews, where other scientists evaluate research methods and results and provide feedback and critique.

Finally, science is a human enterprise, influenced by the values, beliefs, and prejudices of the people who practice it. Researchers should be aware of their biases and strive to minimize their impact on their work. They should also be aware of the broader social, political and economic contexts in which science is practiced and strive to ensure that their work serves the common good.

In summary, science is a systematic and experiential approach to learning about the natural world through observation and experimentation. It is based on empirical evidence, logical thinking, and openness and transparency. Science is a vast field that encompasses many disciplines and is subject to revision and refinement as new evidence emerges. By understanding the principles and methods of science, we can better appreciate the natural world and make informed decisions about the challenges we face as a society.

Science is not just a collection of facts or knowledge, but rather a process of inquiry that involves making observations, formulating hypotheses, and conducting experiments to test those hypotheses. This process requires critical thinking and problem-solving skills, as well as creativity and imagination.

One of the hallmarks of science is its ability to provide explanations for natural phenomena that are consistent with observable evidence. For this reason, scientific discoveries are often referred to as evidence-based, and the scientific method is considered one of the surest ways to understand the world around us.

Science is also interdisciplinary, drawing on principles and techniques from many disciplines to solve complex questions and problems. For example, studying the effects of climate change requires knowledge of geology, atmospheric science, biology and physics, among others.

Another important aspect of science is its ability to create new technologies and innovations that improve our lives. From developing vaccines and antibiotics to creating new materials and energy sources, science has transformed the way we live and work.

However, science is not without its limits and challenges. One of the great challenges facing science today is the problem of repeatability, i. H. The ability to duplicate scientific experiments and discoveries. Many scientific studies have been questioned in recent years due to concerns about the validity of their results, emphasizing the need for greater transparency and rigor in scientific research.

Another challenge facing science is the issue of bias and diversity. Historically, science has been dominated by whites and as a result many perspectives and experiences have been excluded from scientific inquiry. This has led to a lack of diversity in research and unanswered answers to important questions and issues affecting marginalized communities.

Despite these challenges, science remains one of the most important and powerful tools we have to understand the natural world and improve our lives. By developing curiosity, critical thinking, and a deep appreciation for evidence-based inquiry, science education can help prepare students for life in a world increasingly shaped by scientific knowledge and technological innovation.

Chapter No. 2

Scientific Method

The scientific method is a systematic approach to testing hypotheses through observation and experiment. It is the basis of scientific research and is used by scientists from all scientific fields to achieve reliable and valid scientific results.

The scientific method consists of several phases, which include making observations, formulating hypotheses, designing experiments, collecting data, analyzing results, and drawing conclusions. Each step is essential for reliable and valid scientific results.

The first step in the scientific method is observation. Scientists use their senses to observe the natural world and identify patterns or phenomena they wish to study. These observations can be qualitative (descriptive) or quantitative (numerical).

The next step is to make assumptions. A hypothesis is an educated guess about a natural phenomenon that can be tested through experimentation. The assumptions must be verifiable and based on existing scientific knowledge.

Once a hypothesis is formulated, scientists design experiments to test it. The design of the experiment must be carefully planned to ensure that the phenomenon under study can be measured and that other possible factors that could affect the results are eliminated.

The data is then collected as part of the experiment. Data can be quantitative (numerical) or qualitative (descriptive). It is important to collect as much data as possible to ensure the results are accurate and reliable.

After collecting the data, the scientists analyze the results. Statistical analysis is often used to determine the likelihood that results are due to chance or other factors. Researchers also look for patterns or trends in the data that can help them draw conclusions.

Finally, scientists draw conclusions from the results of their experiments. Conclusions should be based on the data collected and the analysis performed.

When a hypothesis is supported by data, scientists can draw broader conclusions or offer new hypotheses for future research.

The scientific method is an iterative process, meaning that it can be iterated with new experiences and data to refine and improve our understanding of natural phenomena. Research results are often published in peer-reviewed scientific journals, where other scientists can review methods, data, and results and provide feedback or suggest improvements.

The scientific method is an essential tool to obtain reliable and valid scientific results. It helps scientists eliminate bias, ensure accuracy, and test hypotheses rigorously and systematically. Understanding the scientific method is an important part of science education and can help students develop critical thinking and problem-solving skills that are useful in all walks of life.

In order to advance the scientific method, it is important to note that this is not a linear process. Rather, it is a cyclical process involving many iterations of hypothesis testing, data collection, analysis, and reasoning. The research process often begins with a question or problem that scientists want to solve. This question or problem leads to the formulation of a hypothesis that can be tested experimentally.

In addition to the steps outlined above, the scientific method also includes a number of other important concepts, such as control groups, independent and dependent variables, and sample size. Control groups are used in experiments to ensure that any observed effects are due to the independent variable being tested and not to other factors. Independent variables are the variables that are manipulated in an experiment, while dependent variables are variables that are measured or observed to determine the effect of the independent variables.

Sample size is also an important factor in scientific research. A larger sample size generally produces more accurate and reliable results. However, larger sample sizes can also be more expensive and take longer to collect, so researchers must weigh the benefits of larger sample sizes against the cost of data collection.

It should also be noted that the scientific method is not infallible. Many factors can affect the results of scientific experiments, such as: b. Experimenter bias, measurement error, and confounding variables. Scientists must consider these factors when designing and conducting experiments to make their results as accurate and reliable as possible.

Despite these limitations, the scientific method remains the most reliable way to test hypotheses and generate scientific knowledge. It has been used for many important discoveries throughout history, such as the discovery of dna and the development of vaccines. The scientific method is also a key tool in solving many of the world's most pressing problems, such as climate change and infectious diseases.

In addition to practical applications, the scientific method also has important philosophical implications. It is based on the principle of empirical evidence, according to which knowledge should be based on observation and experience and not on belief or intuition. This principle has led to many scientific and technological advances and helped shape our understanding of the natural world.

Overall, the scientific method is a powerful tool for producing solid and valid scientific knowledge. By teaching students the scientific method, we can help them develop critical thinking and appreciate the importance of scientific inquiry in our lives.

Chapter No. 3

Observations And Inferences

1. Observations:

Observations are statements of fact based on sensory data. Scientists use their senses, including sight, hearing, smell, touch, and taste, to observe the natural world. Observations can be qualitative or quantitative. Qualitative observations describe the quality of the data, while quantitative observations describe the amount of data. For example, a qualitative observation might be that the plant has green leaves, while a quantitative observation might be that the plant is 20 centimeters tall. Accurate and detailed observations are fundamental to the scientific method as they form the basis for further scientific research.

2. Conclusions:

Conclusions are conclusions drawn from observations. These conclusions can be logical or illogical depending on the quality of the observations and the reasoning used to reach the conclusion. Inferences may also be based on prior knowledge or experience, as well as cultural or social factors. For example, if a scientist notices that a plant is wilting, he can conclude that the plant is not getting enough water. However, it is important to verify this conclusion with other observations and experiments to confirm or refute the conclusion.

3. Differences between observations and conclusions:

It is important to distinguish between observations and conclusions in order to develop sound scientific explanations. Observations are factual and objective, while conclusions are subjective and dependent on the observer's interpretation. Observations are based on sensory data, while conclusions are based on reasoning and interpretation. Conclusions can be influenced by bias or bias, while observations should be made without bias or bias.

4. Importance of observations and conclusions in science:

Observations and conclusions are fundamental to scientific research. Scientists use observations to describe the natural world and form hypotheses, while inference is used to draw conclusions from those observations. Accurate and detailed observations are essential to scientific research and experimentation, and logical deduction helps scientists develop sound scientific explanations. Developing these skills is important for students to become better critical thinkers and problem solvers.

5. Teaching observation and reasoning in science:

Teaching students observation and reasoning is an important part of teaching science. Students must learn to make accurate and detailed observations and draw logical conclusions from those observations. By developing these skills, students can become better observers and critical thinkers. They can also learn to spot and analyze flaws in their own observations and conclusions, leading to more accurate and reliable scientific explanations.

6. Examples of observations and conclusions:

Observations and conclusions can be demonstrated using a variety of scientific phenomena. For example, observing the behavior of animals in their natural environment can provide inferences about their social structure, feeding habits, and reproductive cycles. Likewise, the observation of plant growth can lead to conclusions about the influence of various environmental factors such as light, water and temperature. Students can also practice observation and reasoning in the classroom by conducting experiments, observing natural phenomena, or analyzing scientific research data.

7. Developing observation skills:

Developing observation skills is an important part of science education. Students can improve their observation skills by practicing accurate and detailed observation of the natural world. Teachers can provide opportunities for students to observe live animals, plants, or other natural phenomena. They may also provide aids such as magnifying glasses, binoculars, or microscopes to facilitate observation. Students can also learn to organize their observations using diagrams, drawings, or written notes.

8. Thinking skill development:

Thinking skill development is another important part of science education. Students can learn to make logical inferences by analyzing their observations and looking for patterns or relationships. Teachers can allow students to make predictions based on their observations and then test those predictions with other observations or experiments. Students can also learn to spot errors in their own observations and conclusions, such as b. Confirmation biases, and learn to adjust their thinking accordingly.

9. The importance of critical thinking in observation and conclusion:

Critical thinking is essential to accurate observation and logical thinking. Students should be encouraged to question their assumptions and consider alternative explanations for the phenomena they observe. They must also learn to evaluate the quality of their observations and conclusions, and to recognize when further observations or experiments are needed. Developing critical thinking skills can help students solve problems better and communicate scientific ideas more effectively.

10. Conclusion:

Observations and inferences are fundamental to scientific method and research. By developing the ability to make precise, detailed observations and to draw logical conclusions from those observations, students can become better critical thinkers and problem solvers. These skills are essential for success in science and many other walks of life, and can help students deepen their understanding and appreciation of the natural world.

Chapter No. 4

Tools Of Science

1. Introduction:

Science is an empirical discipline that relies on observation, measurement, and experimentation to understand the natural world. To conduct scientific research, scientists use a variety of tools and techniques that allow them to make precise measurements, collect data, and analyze the results. In this section we look at some of the more common scientific tools.

2. Measuring devices:

One of the most basic tasks in science is measurement. Scientists use a variety of measurement tools to obtain accurate and precise measurements. Some common measuring tools are rulers, scales, thermometers, and graduated cylinders. Each tool is designed for a specific type of measurement, and researchers must choose the right tool for the task at hand.

3. Microscopes:

Microscopes are essential tools for studying the microscopic world. They allow scientists to observe cells, microorganisms and other tiny structures that are not visible to the naked eye. There are different types of microscopes, including light microscopes, electron microscopes, and scanning probe microscopes, each with their own strengths and limitations.

4. Telescopes:

Telescopes are used to observe distant objects such as planets, stars, and galaxies. They collect and focus light from distant objects, allowing scientists to study them in detail. There are different types of telescopes, including refractor telescopes, reflecting telescopes, and radio telescopes, each with their own strengths and limitations.

5. Computers:

computers are indispensable tools for modern scientific research. They are used for data analysis, modeling, simulation and reporting. Scientists use a variety of applications such as spreadsheets, databases, and statistical analysis tools to analyze data and create models of complex systems.

6. Laboratory equipment:

In addition to measuring devices, microscopes, telescopes and computers, scientists use a large number of special devices in their laboratories. This equipment includes, but is not limited to, centrifuges, spectrophotometers, electrophoresis systems, and chromatography columns. Each device is designed for a specific type of experiment or analysis.

7. Safety facilities:

Safety is the top priority in scientific research. Scientists use a variety of safety devices to protect themselves and others in the laboratory from harm. This equipment includes, but is not limited to, goggles, lab coats, gloves and respirators. Researchers are also trained in safe laboratory practices such as chemical handling, equipment use, and waste disposal.

8. Robotics and automation:

In recent years, robotics and automation have become increasingly important tools in scientific research. Robots are used to perform tasks that are too dangerous, difficult, or time-consuming for humans, such as exploring the depths of the ocean, experimenting in space, or analyzing large amounts of data. Automated systems such as high-capacity screens are used to perform repetitive tasks quickly and accurately, saving time and increasing productivity.

9. Imaging tools:

Imaging tools are used to visualize and study the structure and function of biological and non-biological systems. Imaging tools commonly used in science include x-ray machines, magnetic resonance imaging (mri) scanners, computed tomography (ct) scanners and ultrasound machines. Each imaging tool has its strengths and limitations, and scientists must select the right tool for the task at hand.

10. Spectroscopy:

Spectroscopy is the study of the interaction between matter and electromagnetic radiation. It is used to study the properties of atoms and molecules as well as the structure and function of materials. There are many types of spectroscopy, including infrared spectroscopy, ultraviolet light spectroscopy, and nuclear magnetic resonance (nmr) spectroscopy, each with their own strengths and limitations.

11. Data visualization tools:

Data visualization tools are used to create visual representations of scientific data. These tools include but are not limited to graphs, charts, and maps. Data visualization tools enable scientists to quickly and easily spot patterns and trends in their data and share their insights with others.

12. Communication tools:

Communication tools are essential in research. Researchers need to be able to communicate their ideas, methods and results to their peers and the public. Some popular communication tools used in science are scientific journals, conferences, and social media platforms. Effective communication is key to the advancement of science and the dissemination of scientific knowledge.

13. Ethics and integrity:

Ethics and integrity are fundamental elements of scientific research. In their research, scientists must adhere to strict ethical standards, including obtaining informed consent from study participants, ensuring the safety and welfare of humans and animals, and avoiding conflicts of interest. Scientific integrity includes the integrity, objectivity and transparency of research, including the communication of all results, including unexpected or negative ones.

14. Conclusion:

Conclusion: learning tools are diverse and constantly evolving. From measurement tools to microscopes, telescopes, computers, specialized laboratory and safety equipment, robotics and automation, imaging tools, spectroscopy, data visualization tools, communication tools and ethical standards - each tool has its strengths and limitations. Scientists must select the right tools for their task and use them responsibly and ethically. With these tools, scientists can make new discoveries, solve problems, and better understand the world around us.

Chapter No. 5

Properties Of Matter

Matter is anything that has mass and takes up space. Based on its physical properties, it can be classified as solid, liquid or gas. Material properties are properties that can be observed or measured without changing the chemical composition of a substance. These properties include:

1. Mass:

Mass is the amount of matter in an object. It is measured with a scale in grams (g) or kilograms (kg).

2. Volume:

Volume is the amount of space occupied by an object. It is measured in cubic centimeters (cc or cm^3) or liters (l) using a graduated cylinder or other measuring device.

3. Density:

Density is the amount of mass per unit volume. It is a measure of the compactness or distribution of particles in a substance. Density is calculated by dividing an object's mass by its volume. Density units are usually grams per cubic centimeter (g/cm^3) or kilograms per liter (kg/l).

4. Color:

Color is the visible property of an object that reflects certain wavelengths of light and absorbs others. An object's color can be used to identify or distinguish it from other objects.

5. Texture:

Texture is the impression of an object's surface. It can be rough, smooth, bumpy or slippery, among other things.

6. Shape:

Shape is the physical form of an object. It can be described in terms of length, width and height.

7. State:

State refers to the physical form of matter. Matter can exist in solid, liquid or gaseous form depending on the temperature and pressure.

8. Melting point and boiling point:

The melting point is the temperature at which a solid becomes a liquid, while the boiling point is the temperature at which a liquid becomes a gas. These points are specific to each substance and allow its identification.

9. Solubility:

Solubility is the ability of a substance to dissolve in a liquid. It measures the amount of a substance that can be dissolved in a given volume of liquid.

10. Conductivity:

Conductivity is the ability of a substance to conduct electricity or heat. Metals are good conductors of electricity and heat, while nonmetals tend to be poor conductors.

11. Magnetic properties:

Some substances attract magnets and others do not. A substance can be identified based on its magnetic properties.

12. Chemical reactivity:

chemical reactivity is the ability of a substance to undergo chemical transformations. This can be measured by observing how a substance reacts with other substances such as acids or bases.

Understanding the properties of matter is important in science because it allows scientists to identify and classify substances, predict their behavior under different conditions, and design new materials with specific properties.

13. Viscosity:

Viscosity is a measure of the fluid's resistance to flow. It is related to the thickness or viscosity of the liquid. Viscosity is important in fields such as engineering, geology, and biology.

14. Flexibility:

Flexibility is the ability of a material to stretch and return to its original shape. This is important in fields like engineering, physics, and materials science.

15. Hardness:

Hardness is a measure of a material's resistance to denting or scratching. This is important in fields like geology and materials science.

16. Radioactivity:

Radioactivity is the emission of particles or energy from an unstable atomic nucleus. This is important in fields such as nuclear physics and medicine.

17. Flammability:

Flammability is the ability of a substance to burn or catch fire. It is important in fields such as chemistry and fire safety.

18. Toxicity:

toxicity is the degree to which a substance can harm living organisms. It is important in fields such as pharmacology and environmental science.

Scientists use the properties of matter to identify and classify substances, predict their behavior, and develop new materials with specific properties. For example, engineers might use the properties of different metals to design new alloys that are stronger or more corrosion-resistant. Chemists can use the properties of different chemicals to develop new drugs or materials with specific functions. Geologists can use the properties of rocks to identify the types of minerals and ores they contain.

In general, understanding the properties of matter is fundamental to many scientific fields and has important practical applications in fields such as medicine, engineering, and materials science.

Chapter No. 6

States Of Matter

Matter can exist in different states, also called phases. The most common states of matter are solid, liquid and gaseous.

1. Solid:

A solid has a specific shape and volume. The particles of a solid are densely packed and vibrate in place. Solids are generally incompressible and have a constant density. Examples of solids are ice, wood and metal.

2. Liquid:

Liquids have a definite volume but no definite shape. The molecules of a liquid are close together but can move around each other. Liquids can be poured into a container and shaped. They are usually incompressible and have a constant density. Examples of liquids are water, oil and milk.

3. Gases:

Gases have no definite shape or volume. The gas molecules are far apart and move rapidly in all directions. Gases can be compressed and expanded to fill their container. They do not have a constant density. Examples of gases are oxygen, nitrogen and carbon dioxide.

In addition to these three states, there are other states of matter that occur under extreme conditions. These include:

4. Plasma:

Plasma is a state of matter that occurs at very high temperatures or very low pressures. It is similar to a gas but is made up of ionized particles such as electrons and ions. Plasma is the most common state of matter in the universe and is found in stars and lightning.

5. Bose-Einstein condensate:

Bose-Einstein condensate is a state of matter that occurs at extremely low temperatures. It is a collection of atoms that have been cooled to a temperature close to absolute zero and fused into a single quantum state. This state of matter was first predicted by albert Einstein and satyendra nath bose in the 1920s and observed experimentally in 1995.

The study of the states of matter is important in many areas of science, including chemistry, physics, and materials science. Understanding the properties of each state helps scientists explain and predict the behavior of different substances under different conditions. For example, studying the behavior of gases is important in meteorology and atmospheric science, while studying the properties of solids is important in materials science and engineering.

In addition, transitions between states of matter can have important practical applications. For example, changes in water state from ice to liquid and vapor are important to weather patterns and the earth's hydrological cycle. Changes of state can also be used to separate and purify substances, for example in distillation and chromatography techniques used in chemistry.

Chapter No. 7

Changes In Matter

Substantial changes refer to any change in the properties or composition of a substance. There are two main types of changes in matter: physical changes and chemical changes.

Physical changes are those that affect the physical properties of matter, such as shape, size, and condition, without altering its chemical composition. Physical changes do not create new substances, they only modify existing ones. Examples of physical changes include changes in state (solid, liquid, gas), shape changes, size changes, and phase changes.

Chemical transformations consist in changing the chemical composition of matter. In a chemical transformation, one or more substances transform into new substances with different properties. Chemical changes usually involve the breaking and forming of chemical bonds, resulting in a rearrangement of atoms or molecules. Examples of chemical changes are burning, rusting and etching.

Besides physical and chemical changes, other types of changes in matter can also occur, such as nuclear changes. Nuclear changes include changes in the nucleus of an atom, such as radioactive decay or nuclear fusion, and can lead to the formation of new elements or isotopes.

Understanding how matter changes is essential for many scientific and technological applications. In chemistry, for example, understanding the changes that occur in chemical reactions is essential to designing new materials and understanding the properties of substances. In engineering and materials science, understanding the physical changes that occur in materials can help design and optimize materials for specific applications.

In addition, in many everyday applications such as cooking, cleaning and manufacturing, it is important to understand how matter changes. In the kitchen, for example, understanding the physical and chemical changes that occur during the cooking process is critical to preparing tasty and safe meals. In manufacturing, understanding changes in matter is important for the design and production of high value products such as electronics and pharmaceuticals.

In short, a change of substance refers to any change in the properties or composition of a substance. Physical changes involve changes in the physical properties of matter, while chemical changes involve a change in the chemical composition of matter. Understanding the changes occurring in matter is essential for many scientific and technological applications, as well as for everyday activities.

Understanding change in matter is a fundamental concept in science and plays an important role in many fields, including chemistry, physics, and materials science. Physical changes fall into four main categories: changes in state, changes in shape, changes in size, and changes in phase.

Changes of state occur when a substance changes from one physical state to another, for example from a solid to a liquid or from a liquid to a gas. It is caused by a change in temperature or pressure. For example, when ice is heated, it melts and turns into water. When water is heated, it evaporates and turns into steam.

Shape changes occur when a substance changes shape but not volume or mass. For example, when a piece of clay is shaped into a different shape, it undergoes a physical change in shape. However, the amount of sound remains the same.

Size changes occur when the volume of a substance changes but its mass and composition remain the same. For example, if a sheet of paper is torn into small pieces, the overall weight of the paper will remain the same, but the size and shape of the paper will change.

Phase changes occur when a substance changes from one phase to another. A phase is a region of matter with uniform physical and chemical properties. The most common phases of matter are solid, liquid and gas. There are also other phases such as plasma and bose-einstein condensates. For example, when water is frozen it changes from a liquid phase to a solid phase, and when it is boiled it changes from a liquid phase to a gaseous phase.

Chemical transformations, in turn, consist of breaking and forming chemical bonds between atoms or molecules, resulting in a change in the composition of a substance. In a chemical conversion, the starting materials react to form new substances with different properties. Examples of chemical changes are combustion, rust and fermentation.

One of the most important aspects of metabolism is weight maintenance. The law of conservation of mass states that the total mass of the reactants in a chemical reaction equals the total mass of the products. This means that the mass of a substance remains constant when there are physical or chemical changes, even if its properties may change.

Understanding how matter changes is essential for many scientific and technological applications. In chemistry, understanding chemical transformations is essential for the design and synthesis of new materials and for understanding the properties of substances. In physics, understanding phase changes is important to understand how materials behave under different conditions. In materials science, understanding the physical changes in materials is important to design and optimize materials for specific applications.

In summary, substance changes are basic scientific concepts that involve changes in the properties or composition of a substance. Physical changes involve changes in the physical properties of matter, while chemical changes involve a change in the chemical composition of matter. Understanding how matter changes is essential for many scientific and technological applications and plays an important role in many scientific fields, including chemistry, physics and materials science.

Chapter No. 8

Introduction To Energy

Energy is a fundamental scientific term that refers to the ability of a system to function. It is essential to life and is involved in all physical and chemical processes. Understanding energy and its properties is important for many scientific and technological applications, including power generation, environmental science, and materials science.

Energy comes in many forms, including kinetic energy, potential energy, thermal energy, electrical energy, and chemical energy, to name a few. Kinetic energy is the energy of motion while potential energy is the energy stored in an object due to its position or configuration. Thermal energy is the energy related to an object's temperature, while electrical energy is the energy carried by the flow of electrical charges. Chemical energy is the energy stored in the chemical bonds between atoms or molecules.

One of the most important energy principles is the law of conservation of energy, which states that energy can neither be created nor destroyed, but can only be converted from one form to another. This means that the total amount of energy in the system remains constant, although it can be converted into different forms.

The unit of energy measurement is the joule (j), although other units such as calories or electron volts can be used. Energy can be transferred between systems by various means such as heat, work, or radiation. For example, when a person eats food, the body converts the chemical energy stored in the food into heat energy and kinetic energy that can be used to do work.

Understanding energy is essential to many scientific and technological applications. In power generation, understanding the properties of different forms of energy is essential for the design and optimization of energy sources such as solar panels and wind turbines. In environmental science, understanding the energy balance of different ecosystems is important to understand the impact of human activities on the environment. In materials science, understanding the properties of materials at the atomic and molecular level is essential for the design and optimization of new materials for various applications.

Energy is a fundamental scientific concept that relates to a system's ability to do work. It comes in many forms and is involved in all physical and chemical processes. Understanding energy and its properties is essential for many scientific and technological applications, including power generation, environmental science, and materials science. The law of conservation of energy, which states that energy cannot be created or destroyed, only transformed from one form to another, is a fundamental energy principle that plays an important role in many scientific fields.

Energy is a key concept in many scientific fields, including physics, chemistry and biology. In physics, energy is described as the ability of a physical system to do work, which is defined as the product of the force acting on an object and the distance the object travels in the direction of the force. . Work is a fundamental concept in physics, and understanding energy is key to understanding how systems behave and interact.

An important aspect of energy is the distinction between potential energy and kinetic energy. Potential energy is the energy stored in an object due to its position or configuration, while kinetic energy is the energy of motion. For example, an object on a hill has potential energy due to its position, while an object in motion has kinetic energy due to its motion.

Another important aspect of energy is its different forms. In addition to kinetic and potential energy, energy can be thermal, electrical, chemical or nuclear, among others. Thermal energy is the energy related to the temperature of an object or system, while electrical energy is the energy carried by the flow of electrical charges. Chemical energy is the energy stored in the chemical bonds between atoms or molecules, while nuclear energy is the energy released from the nuclei of atoms during nuclear reactions.

Understanding energy is also important to understanding the relationship between matter and energy. According to einstein's famous equation $e=mc^2$, matter and energy are interchangeable and interchangeable. This has important implications for many areas of science, including nuclear energy and particle physics.

In addition to the different forms of energy, scientists also study the transfer of energy between systems. Energy can be transferred between systems by various means such as heat, work, or radiation. For example, when a stove heats a pot of water, thermal energy is transferred from the stove to the water through heat exchange. Similarly, when a battery powers an electrical device, power is transferred from the battery to the device.

Finally, understanding energy is important for many real-world applications. In addition to power generation and environmental science, energy is also an important driver in industries such as transport, medicine and telecommunications. Understanding energy and its properties is therefore essential in many areas of scientific research and technological development.

Energy is a basic scientific term that refers to the ability of a physical system to do work. Understanding energy is key to understanding the behavior and interaction of systems, as well as many real-world applications. Energy can take many forms and can be transferred between systems in different ways. Scientists continue to study energy and its properties, and new discoveries in this field could revolutionize many areas of science and technology.

An important area of energy-related research is thermodynamics, i.e. The part of physics that deals with the relationships between heat, temperature, energy and work. Thermodynamics helps scientists understand how energy is transferred and converted between different systems and has many practical applications, including the design of engines, cooling systems, and power plants.

Another important area of energy-related research is renewable energy, which refers to energy sources that naturally recharge and do not deplete over time. Examples of renewable energy sources are solar, wind, hydroelectric and geothermal. Renewable energy is becoming increasingly important as concerns about climate change and energy security continue to grow. Scientists and engineers are working on the development of new and improved technologies for the use of renewable energy sources and their integration into the existing energy infrastructure.

In addition to scientific and technological applications, energy also has important social and economic implications. Access to energy is central to many aspects of

modern life, including transportation, communications and healthcare. However, access to energy is not evenly distributed around the world and many people, especially in developing countries, do not have access to reliable and affordable sources of energy. Energy production and consumption also have significant environmental impacts on the
, including air and water pollution and greenhouse gas emissions, which can contribute to climate change and other environmental problems.

Finally, understanding energy requires a multidisciplinary approach that includes knowledge of physics, chemistry, biology and engineering, among others. Scientists and engineers in these disciplines work together to study energy and its properties, to develop new technologies for energy production and use, and to address the many challenges and opportunities associated with energy production and consumption.

In summary, energy is a fundamental scientific concept with many different forms and properties. Understanding energy is essential to understanding the behavior and interactions of physical systems, and has many scientific, technological, social, and economic implications. Scientists and engineers continue to research energy and develop new technologies to use it in a sustainable and responsible manner.

Chapter No. 9

Light And Sound

Light and sound are two important forms of energy that are essential in many areas of our daily lives. Light is a type of electromagnetic radiation that can be perceived by the human eye, while sound is a mechanical wave that propagates through a medium such as air, water, or solids.

Light is an important source of information about the world around us. It allows us to see colors, shapes and textures and is essential for many daily activities such as reading, driving and navigating our surroundings. Light also plays an important role in scientific research, with applications in fields such as astronomy, biology, and physics.

One of the important properties of light is that it can be broken down into different colors or wavelengths, as seen in the rainbow. This property has many practical applications, including in lighting design and display technology. Another important property of light is that it can be reflected, refracted or absorbed by different materials, which allows us to see objects and images.

On the other hand, sound is an important form of energy for communication and entertainment. It allows us to hear words, music, and other sounds and plays a key role in human interaction and socialization. Sound also has many scientific and technological applications, particularly in fields such as acoustics, audiology, and sound engineering.

One of the important properties of sound is that it can be described in terms of frequency and amplitude. Frequency refers to the number of cycles per second, or hertz (hz), while amplitude refers to the strength or intensity of the sound wave. These properties determine how we perceive sounds, including their pitch, volume, and timbre.

Light and sound have many practical applications in science and technology. In medicine, for example, light is used in imaging processes such as x-rays, ct scans and mris, and sound in diagnostic processes such as ultrasound. In communication and entertainment, light and sound are used to transmit and transmit information, especially in radio, television and the internet.

In addition to practical applications, light and sound also have many artistic and cultural implications. They play an important role in music, theatre, film and other artistic expressions and have been used throughout history to convey emotion, mood and meaning.

Light:

one of the important properties of light is that it can behave both as a wave and as a particle, depending on the context of observation. This is called wave-particle duality and is a fundamental concept of quantum mechanics. It explains many of the strange and seemingly paradoxical properties of light, such as its ability to penetrate solids (as in x-rays) and its ability to create interference patterns (as in the famous double-slit experiment).

Another important property of light is that it travels at a constant speed of about 299,792 kilometers per second (or 186,282 miles per second) in a vacuum. This is called the speed of light and is the maximum speed at which anything in the universe can travel. The speed of light is also important for understanding the concept of time dilation in einstein's theory of relativity.

Light can also be polarized, meaning that its electric and magnetic fields point in a certain direction. This property is used in many technologies, including sunglasses, lcd screens, and 3d films.

Sound:

one of the important properties of sound is that it can be described in terms of wavelength and frequency. Wavelength is the distance between two adjacent peaks or troughs of a sound wave, while frequency is the number of waves that pass through a point per second. These properties determine pitch and timbre and can be measured with special equipment such as an oscilloscope.

Another important property of sound is that it can be reflected, refracted, or scattered by various materials, allowing it to be manipulated and directed. This property is used in many technologies, including microphones, speakers, and hearing aids.

Sound can also be used in non-invasive medical imaging such as ultrasound. It involves sending high-frequency sound waves through the body and measuring their reflections to create images of internal organs and tissues. Ultrasound is widely used in prenatal care because it can provide detailed images of the developing fetus without exposing the mother or baby to harmful radiation.

Applications:

light and sound have many practical applications in science and technology. In astronomy, for example, light is used to study the properties and behavior of stars, galaxies, and other celestial bodies. Telescopes and other imaging equipment are used to capture images of these objects at different wavelengths of light, allowing astronomers to learn more about their composition, temperature, and other.

Properties.

In medicine, light and sound are used in a variety of diagnostic and therapeutic procedures. As mentioned above, light is used in imaging procedures such as x-rays, ct scans and mris, while sound is used in diagnostic procedures such as ultrasound. The light is also used in laser surgery, where high-intensity light is used to remove or destroy diseased tissue.

In communication and entertainment, light and sound are used to transmit and transmit information, including radio, television and the internet. Light is also used to transmit data in optical fibers, which are thin, flexible cables that can carry information over long distances with minimal loss of signal quality.

In addition to practical applications, light and sound also have many artistic and cultural implications. Used in music, theater, film, and other artistic expressions, they can be used to convey emotion, mood, and meaning. For example, in music, different combinations of pitch, timbre, and rhythm can evoke different emotions and create different moods.

finally, the study of light and sound requires a multidisciplinary approach that includes knowledge of physics, chemistry, biology and engineering, among others. Scientists and engineers in these disciplines work together to study the properties of light and sound, develop new technologies to harness and manipulate these forms of energy, and explore the many practical and artistic uses of light and sound.

Light and sound are two important forms of energy with many different properties and uses. They are essential to many aspects of our daily lives, including communication, entertainment and research, and have important social, cultural and artistic implications.

Light and sound are fascinating phenomena that have been studied and understood for centuries. Let's take a closer look at some of the properties and uses of light and sound.

Chapter No. 10

Forces And Motion

Forces and motion are fundamental concepts in physics that are essential to understanding how objects behave in the natural world. Simply put, motion refers to the movement of an object from one place to another, while force is a push or pull that can cause movement or change the direction and speed of a moving object. .

According to newton's laws of motion, an object remains stationary or moves smoothly in a straight line unless acted upon by an external force. That is, if no force is applied to the object, it will continue to move at a constant speed and in a straight line. However, when a force is applied, the object accelerates or changes direction.

Force can be described in various units of measurement, such as newton (n) or pounds (pounds), and can be broken down into several types, including gravitational, electromagnetic, nuclear, and frictional. For example, gravitational force is the attraction between two massive objects like the earth and the moon, while electromagnetic force is the force that holds atoms and molecules together.

In addition to force, other factors can affect an object's motion, including the object's mass, the amount of force applied, and the presence of friction or drag. Mass refers to the amount of matter in an object and is usually measured in kilograms (kg) or grams (g). The greater the mass of an object, the greater the force required to accelerate it.

Friction, on the other hand, is a force that resists movement between two touching surfaces and can cause an object to slow down or stop. Air resistance is a type of friction that occurs when an object moves through air, and it can also affect an object's movement by slowing it down.

Understanding the relationship between forces and motion is critical in many fields, including engineering, mechanics, and physics. By studying the principles of force and motion, scientists and engineers can develop new technologies and materials that can help us better understand and control the natural world.

One of the important concepts related to forces and motion is the concept of acceleration. Acceleration is the rate at which an object's velocity (velocity and direction) changes over time. It is directly proportional to the force exerted on the object and inversely proportional to the mass of the object. Mathematically, the acceleration can be expressed as $a = f/m$, where "a" is the acceleration "f"; is the force and "m" is the mass.

Another important concept related to forces is momentum. Momentum is the product of an object's mass and velocity and represents the object's momentum. The momentum of the object is conserved in the absence of external forces. That is, when two objects collide, their total momentum before the collision is equal to their total momentum after the collision, regardless of the forces that were acting during the collision.

There are also different types of forces acting on objects. Gravity is the gravitational pull between two massive objects and is responsible for holding planets and stars in orbit. The electromagnetic force is the force that governs the interactions between electrically charged particles and is responsible for phenomena such as electricity and magnetism. The nuclear force is the force that holds atomic nuclei together and is responsible for the energy released in nuclear reactions. Friction is the force opposing movement between two contact surfaces and is responsible for phenomena such as tire friction and air resistance.

Forces and motion play a key role in many real-world applications, from designing cars and airplanes to predicting a satellite's orbit through space. Engineers and physicists use the principles of force and motion to design safe, efficient, and reliable machines and structures. In addition, understanding the principles of force and motion can help us understand the behavior of natural phenomena such as the movement of planets and stars, the behavior of waves and tides, and the dynamics of weather patterns.

One of the most important concepts related to forces and motion is work. Work is done when a force causes an object to move a distance and is defined as the product of force and displacement. Mathematically, work can be expressed as $w = f \times d$, where "w" is work, "f"; is the force and "d" is the displacement. The unit of work is the joule (j).

Force is another important concept related to forces and motion. Power is the speed at which work is done and is defined as the work done per unit time. Mathematically, power can be expressed as $p = w/t$, where "p" is power "w"; is work and "t" is time. The unit of power is the watt (w).

Another important concept related to forces and motion is energy. Energy is the ability to do work and can be divided into two main types: kinetic energy and potential energy. Kinetic energy is the energy an object possesses as a result of its motion and is directly proportional to its mass and the square of its speed. Potential energy is the energy an object possesses due to its position or configuration and can be gravitational, elastic or chemical in nature.

Finally, understanding the principles of force and motion is essential to many practical applications, from designing roller coasters and amusement park rides to developing new materials and technologies for space exploration. In addition, the principles of forces and motion are fundamental to many other branches of science, including chemistry, biology, and geology, and can help us understand the behavior of complex natural systems such as the human body and the earth's planetary system.

Chapter No. 11

Magnetism

Magnetism is a phenomenon that occurs when certain materials attract or repel other materials. This is due to the presence of a magnetic field, which is a force field surrounding a moving magnet or electrically charged object. The study of magnetism is an important part of physics and has many practical applications, including electronics, medicine, and power generation.

One of the main properties of magnetism is polarity. Magnetic objects have two poles, the north pole and the south pole. Like poles (e.g. Two north or two south poles) repel each other, while opposite poles (e.g. North and south poles) attract. This is evident in the behavior of magnets, which can be used to lift and move other magnetic objects.

Magnetic fields are created by the movement of electrical charges. It is called the electromagnetic force and is one of the four fundamental forces of nature along with gravity, the strong nuclear force and the weak nuclear force. For example, when an electric current flows through a wire, it creates a magnetic field around it. It is the basis of many electrical devices including motors, generators and transformers.

One of the most important concepts related to magnetism is induction. Induction occurs when a changing magnetic field induces an electric current in a nearby conductor. It is the basis of many modern technologies, including power generation and wireless communications.

Another important concept related to magnetism is magnetic field strength, which is the force exerted by a magnetic field on a magnetic object. The strength of the magnetic field depends on the distance to the object and the strength of the magnetic object creating the field. Finally, an understanding of the principles of magnetism is essential to many practical applications, including the medical field, where magnetic resonance imaging (mri) is used to visualize internal organs and tissues, and the field of power generation, where magnetic fields are used to generate and distribute electricity. In addition, understanding the principles of magnetism can help us understand the behavior of natural phenomena such as the earth's magnetic field and the behavior of stars and galaxies in space.

Magnetic materials: not all materials are magnetic, but some materials are strongly attracted to magnets or can become magnets themselves when exposed to a magnetic field. These materials are called ferromagnetic materials and include iron, nickel and cobalt. Other materials such as aluminum and copper are not strongly magnetic but can still be affected by magnetic fields.

Magnetic domains: magnetic materials are made up of tiny regions called magnetic domains, which are composed of aligned magnetic moments of the atoms. In an magnetized material, these domains are randomly aligned and cancel each other out. But when the material is magnetized, the domains align with the magnetic field, creating a net magnetic field.

Earth's magnetic field: the earth has a magnetic field similar to that of a bar magnet, with north and south poles. The magnetic field is believed to be created by the movement of molten iron in the earth's outer core.

Electromagnets: electromagnets are magnets that are created by passing an electric current through a wire. The magnetic field created by the current can be increased or decreased by changing the amperage, making electromagnets very versatile and useful in a variety of applications.

Magnetic and electric fields: magnetic and electric fields are closely related and both are part of the electromagnetic force. A changing magnetic field can induce an electric field and vice versa. It is the basis for electromagnetic waves, such as b. Radio waves used for communication and information transmission.

Applications of magnetism: magnetism has many practical applications, including medicine, power generation, transportation, and information technology. Magnetic materials are used in electric motors, generators, and transformers, while electromagnets are used in a variety of applications including mri machines, maglev trains, and particle accelerators.

Magnetic hysteresis: if a magnetic material is magnetized and then demagnetized, it may not return to its original magnetic state. This is called magnetic hysteresis and is due to the fact that the magnetic domains in the material do not always return to their original alignment. This effect is important when designing magnetic materials for specific applications.

Maglev: magnetic levitation or magnetic levitation is a technology that uses magnetic fields to lift and propel objects without physical contact. Maglev trains, for example, use powerful electromagnets to lift the train off the tracks and propel it forward. This technology has the potential to revolutionize transportation by allowing you to travel at high speeds with minimal friction and noise.

Superconductivity: some materials, so-called superconductors, can conduct electricity without resistance when cooled to very low temperatures. Superconductors can also emit magnetic fields, a phenomenon known as the meissner effect. This makes them useful in many applications, including mri machines and particle accelerators.

Magnetic monopoles: while all magnets have north and south poles, it is theoretically possible that there are particles with only one pole, called magnetic monopoles. Although magnetic monopoles have not yet been observed, some particle physics theories predict their existence.

Magnetic reconnection: magnetic reconnection occurs when magnetic field lines in the plasma (a gas of charged particles) break and then reconnect, releasing large amounts of energy. This process is believed to be responsible for various natural phenomena, including solar flares and the northern lights.

Magnetic field measurement: magnetic fields can be measured using a variety of instruments, including magnetometers and gauss meters. Measuring the strength and direction of the magnetic field at a given location, these instruments are important tools for a variety of applications, including geology and environmental science.

In summary, magnetism is a complex and multifaceted phenomenon that is fundamental to our understanding of the natural world and has many practical applications. From the behavior of magnets to the design of magnetic materials for specific applications, understanding the principles of magnetism is essential in many areas of science and technology.

Chapter No. 12

Electricity

Electricity is a fundamental aspect of modern life, powering everything from the lights in our homes to the appliances we use every day. Here are some key concepts related to electricity:

electric charges: all matter is made up of atoms, which include positively charged protons, negatively charged electrons, and neutral neutrons. When electrons move from one atom to another, they create an electrical charge. If an object has an excess of electrons, it is negatively charged; if it lacks electrons, it is positively charged.

Electric fields: electric charges create electric fields that can exert a force on other charged objects. The strength of the electric field is determined by the distance between charged objects and the magnitude of the charges.

Electric currents: when a group of charged particles, such as electrons, move together in a certain direction, an electric current is generated. Electric currents can be direct current (dc) flowing in one direction or alternating current (ac) changing direction.

Resistance: resistance is a measure of a material's resistance to the flow of electric current. High-strength materials such as rubber and plastics do not conduct electricity well, while low-strength materials such as metals conduct electricity well.

Ohm's law: ohm's law states that the current flowing through a conductor between two points is directly proportional to the voltage at those two points and inversely proportional to the resistance between them. This relationship is often expressed by the equation $i = v/r$, where i is current, v is voltage, and r is resistance.

Circuits: a circuit is a closed path through which electric current can flow. Circuits can be as simple as a single light bulb connected to a battery or as complex as the wiring in a house.

Electromagnetism: electromagnetism is the interaction between electric and magnetic fields. Moving electric charges create magnetic fields, while changing magnetic fields create electric fields. This relationship underlies many important technologies such as electric motors and generators.

Capacitance: capacitance is a measure of a material's ability to store an electrical charge. Comprised of two conductive plates separated by an insulating material, capacitors are used in various applications to store and release electrical energy.

Inductance: inductance is a measure of the degree to which a changing magnetic field induces an electric current in a circuit. This effect underlies many important technologies such as transformers and electrical generators.

Electrical safety: electricity can be dangerous so always take proper precautions when working with electrical circuits or equipment. This includes using insulated tools, turning off power sources before working on electrical circuits, and wearing protective clothing and equipment as needed.

Electrical potential: electrical potential, also called voltage, is a measure of electrical potential energy per unit charge in an electrical circuit. The greater the electric potential difference between two points in a circuit, the higher the voltage and the stronger the electric field between them.

Power: power is the speed at which energy is transferred or converted. In electrical circuits, power is measured in watts (w) and is equal to the product of voltage and current.

Series and parallel circuits :

In a series circuit, components are connected in sequence so that current flows through each component in turn. In a parallel circuit, components are connected together, allowing current to flow through each component independently.

Diodes: diodes are electronic devices that allow current to flow in one direction but not the other. They are commonly used in rectifiers and voltage regulators.

Transistors: transistors are electronic devices that can be used as switches or amplifiers. They are commonly used in digital circuits such as those found in computers and other electronic devices.

Integrated circuits: integrated circuits, also known as microchips, are electronic circuits etched into a single piece of semiconductor material. They are used in a variety of electronic devices, from digital calculators and watches to smartphones and computers.

Electric fields and potential energy: electric fields have the ability to store potential energy that can be released when charges are allowed to move. Capacitors are devices that can store electrical charge and slowly release it over time, providing a power source for various applications.

Electricity generation: electricity can be generated in a variety of ways, including fossil fuels, nuclear power, hydroelectric power, and renewable sources such as solar and wind power. The conversion of energy from one form to another, such as converting mechanical energy to electrical energy in a generator, is a key concept in electrical engineering.

These are just a few of the many concepts related to electricity. Understanding these principles is critical to a variety of applications, from the design of electrical systems to the development of new technologies.

In general, electricity is a fundamental aspect of modern life and understanding its principles is essential in a variety of fields including engineering, physics and technology. From the behavior of electric charges to the design of electric circuits to precautions in handling electricity, the study of electricity is at the heart of our daily lives.

Chapter No. 13

Plants And Animals

Plants and animals are the two main groups of living organisms on earth. They differ in physical characteristics, behavior and physiology. Here are some concepts related to plants and animals:

cell structure: plants and animals are made up of cells, but their cell structure is different. Unlike animal cells, plant cells have a rigid cellulose cell wall. Unlike animal cells, plant cells also contain chloroplasts, which are responsible for photosynthesis.

Photosynthesis: photosynthesis is the process by which plants convert sunlight into energy in the form of glucose. During photosynthesis, plants use carbon dioxide and water to produce oxygen and glucose.

Respiration: respiration is the process by which living organisms break down glucose to release energy. In animals, respiration occurs through the lungs, in plants through small openings in the leaves called stomata.

Reproduction: plants and animals reproduce in different ways. Most plants reproduce via seeds or spores, while animals reproduce via eggs or live births.

Adaptation: plants and animals have evolved adaptations that help them survive in their environment. For example, plants have evolved spines or toxic chemicals to deter herbivores, while animals have evolved camouflage or defensive behaviors to evade predators.

Ecosystems: plants and animals interact in complex ecosystems. Plants provide animals with food and oxygen, while animals help pollinate plants and disperse their seeds.

Taxonomy: taxonomy is the science of naming and classifying living organisms. Plants and animals are divided into several categories based on their physical characteristics and genetic makeup.

Endangered species: many animal and plant species are threatened by habitat destruction, pollution and other human activities. Conservation measures aim to preserve these species and their habitats.

Anatomy: the anatomy of plants and animals is very different. For example, plants have roots, stems, and leaves, while animals have organs like the heart, lungs, and brain.

Behavior: plants and animals also show different behaviors. For example, animals can move and display social behaviors, while plants are usually immobile and do not display social behaviors.

Homeostasis: homeostasis is the process by which living organisms maintain a stable internal environment. Plants and animals use homeostasis to regulate processes such as temperature, ph, and water balance.

Genetics: genetics is the science of inheritance of traits from generation to generation. Plants and animals have unique genetic codes that determine their physical characteristics and behavior.

Evolution: evolution is the process by which living organisms change over time through natural selection. Plants and animals have evolved to adapt to changing environmental conditions and compete for resources.

Nutrition: plants and animals need nutrients to survive and grow. Plants get nutrients through their roots and leaves, while animals get nutrients through their diet. Immune system: plants and animals have an immune system that protects them from pathogens such as viruses and bacteria. Plants use different strategies including chemical defenses and physical barriers, while animals have specialized cells and organs such as white blood cells and lymph nodes.

Migration: migration is the seasonal movement of animals from one region to another. Many birds, fish, and mammals migrate for food resources and nesting opportunities.

Fossil record: plant and animal fossils provide important information about the history of life on earth. Fossils can provide information about how species evolved over time and how they adapted to changing environmental conditions.

Biotechnology: biotechnology is the use of living organisms or their components to develop new technologies and products. Plants and animals are used in biotechnology for applications such as genetic engineering and drug development.

Understanding these concepts is essential in many fields, including biology, ecology, and agriculture. Plants and animals play an important role in our lives and in nature, and studying them can help us better understand the world around us.

Plants and animals are diverse and complex living organisms that play important roles in nature and human society. Understanding their biology and behavior is vital in fields such as agriculture, conservation, medicine, and biotechnology.

Chapter No. 14

Habitats And Ecosystems

Habitats and ecosystems are two important concepts in ecology, the study of the interactions between living organisms and their environment. Here are some key concepts related to habitats and ecosystems:

habitat: habitat is the natural environment in which a species lives. Habitats can range from deserts to rainforests, providing the resources necessary for species survival, such as food, water, and shelter.

Biodiversity: biodiversity refers to the variety of life forms in an ecosystem. Habitats with high biodiversity have a large variety of different species, while habitats with low biodiversity have fewer species.

Keystone species: keystone species are species that have a disproportionate impact on an ecosystem relative to their abundance. For example, predators like wolves can help keep an ecosystem in balance by controlling populations of other species.

Trophic levels: trophic levels are the different levels of the food chain or web. Producers (like plants) are at the bottom of the food chain, followed by primary consumers (like herbivores), secondary consumers (like carnivores that eat herbivores), etc.

Ecosystem services: ecosystem services are the benefits that humans derive from ecosystems. These include things like clean water, clean air, and pollination, as well as more tangible resources like wood and food.

Biomes: biomes are large-scale ecosystems characterized by specific climate patterns and vegetation types. Examples of biomes are tundra, tropical rainforests, and grasslands.

Ecological succession: ecological succession is the process by which an ecosystem changes over time. For example, after a disturbance such as a forest fire, new plant species can colonize an area and change the composition of the ecosystem.

Human impact: human activities such as deforestation, pollution and climate change can have significant impacts on habitats and ecosystems. These impacts can lead to biodiversity loss, ecosystem degradation and even ecosystem collapse.

Restoration: restoration is the process of repairing or restoring damaged ecosystems. This may include actions such as reforestation, wetland restoration or habitat creation for endangered species.

Monitoring and assessment: monitoring and assessment are important tools to understand the state and functioning of ecosystems. These activities can include measuring things like biodiversity, nutrient levels, and ecosystem productivity.

Habitat fragmentation: habitat fragmentation is the process by which habitats break up into smaller, more isolated areas. This can occur as a result of human activities such as urbanization or agriculture, leading to a reduction in biodiversity and an increased risk of extinction for some species.

Ecological niches: ecological niches refer to the role that a species plays in its ecosystem, including its interactions with other species and the resources it uses. Understanding the niches of different species can help ecologists predict how they will respond to environmental changes.

Ecological footprint: the ecological footprint is a measure of the impact of human activities on the environment. It covers factors such as energy use, transportation and food production and can help individuals and organizations assess their impact on the environment.

Food webs: a food web is a complex network of interactions between different species in an ecosystem. It shows how energy and nutrients flow from one organism to another and how different species are connected.

Ecosystem resilience: the ecosystem resilience describes the ability of an ecosystem to recover from disturbances such as drought, fire or floods. Resilient ecosystems are better able to maintain their function and structure in the face of environmental change.

Habitat restoration: habitat restoration includes restoring degraded or destroyed habitats to their original condition.this may include actions such as planting native species, removing invasive species, or restoring wetlands.

Ecosystem management: ecosystem management is the process of managing natural resources in a way that balances conservation and human use. This may include activities such as wildlife management, logging and sustainable agriculture.

Ecotourism: ecotourism is a form of tourism that emphasizes visiting natural areas and learning about the ecology and conservation of the region. It can bring economic benefits to local communities while promoting the conservation and sustainable use of natural resources.

Understanding habitats and ecosystems is essential to maintaining the health and functioning of the natural world and ensuring the sustainable development of human societies. By studying these concepts and implementing effective management and conservation strategies, we can help protect the biodiversity and ecological integrity of our planet for future generations.

Chapter No. 15

Food Chains And Webs

Food chains and food webs are important concepts in ecology that describe how energy and nutrients move through ecosystems. Here are some key concepts related to food chains and food webs:

trophic levels: trophic levels refer to different levels of a food chain or web. Producers (like plants) are at the bottom of the food chain, followed by primary consumers (like herbivores), secondary consumers (like carnivores that eat herbivores), etc.

Food chains: a food chain is a linear representation of the flow of energy and nutrients through an ecosystem. It shows the transfer of energy from one trophic level to another, with each organism serving as food for the next level.

Food webs: a food web is a more complex representation of the flow of energy and nutrients through an ecosystem. It shows the interdependence of different food chains and how different species are related to each other through their interactions.

Energy transfer: energy is transferred between trophic levels when organisms are consumed by other organisms. However, not all of the energy from a given trophy level is transferred to the next level. Some of the energy is lost as heat and some is used by the body for growth and metabolism.

Biomass: biomass refers to the total amount of living matter at a given trophic level. Biomass tends to decrease at higher trophic levels as energy is lost as it moves up the food chain.

Eco-efficiency: eco-efficiency refers to the amount of energy transferred between trophic levels. In general, eco-efficiency is low and only a small percentage of energy is transferred from one level to another.

Best carnivores: the best carnivores are organisms that occupy the highest trophic levels of an ecosystem. They have a significant impact on the ecosystem as they regulate the populations of other species.

Keystone species: keystone species are species that have a disproportionate impact on an ecosystem relative to their abundance. For example, predators like wolves can help keep an ecosystem in balance by controlling populations of other species.

Omnivores: omnivores are organisms that feed on both plants and animals. Depending on what they consume, they can occupy different trophic levels.

Human impact: human activities such as overfishing, habitat destruction and climate change can have significant impacts on food chains and webs. These impacts can lead to the collapse of ecosystems and loss of biodiversity.

Primary productivity: primary productivity refers to the rate at which producers (e.g. Plants) produce organic material through photosynthesis. Primary productivity is the basis of all food chains and webs as it provides energy and nutrients to all other organisms in the ecosystem.

Decomposers: decomposers are organisms that break down dead organic matter into inorganic nutrients. They play an important role in nutrient recycling in ecosystems as they release nutrients into the soil or water for plants and other organisms to take up.

Trophy cascades: trophy cascades occur when a change in one trophic level affects many other trophic levels in an ecosystem. For example, removing apex predators can lead to an increase in their prey population, which in turn can lead to a decrease in their prey population.

Biodiversity: biodiversity refers to the variety of living organisms in an ecosystem. Food chains and webs are an essential part of biodiversity as they represent the interactions between different species in an ecosystem.

Invasive species: invasive species are alien species that are introduced into an ecosystem and can damage native species and the ecosystem as a whole. Invasive species can disrupt food chains and webs by competing with native species for resources and introducing new predators or prey.

Ecological niches: ecological niches refer to the role that an organism plays in an ecosystem, including its interactions with other species and resource use. Understanding ecological niches is important to understand the dynamics of food chains and webs, as different species occupy different niches and interact with each other in complex ways.

Understanding food chains and webs is essential to understanding ecosystem dynamics and interactions between different species. By studying these concepts, ecologists can develop strategies to protect biodiversity, manage natural resources sustainably, and reduce the impact of human activities on the environment.

In general, food chains and webs are complex and dynamic systems that play a key role in the functioning of ecosystems. Understanding the relationships between different species and the flow of energy and nutrients through these systems is critical to managing and conserving natural resources and reducing the impact of human activities on the environment.

Chapter No. 16

Life Cycles

Here are some additional terms related to life cycles:

metamorphosis: metamorphosis is a process of transformation that some organisms undergo during their life cycle. For example, metamorphosis in insects includes a number of different stages such as the larval stage, the pupal stage and the adult stage.

Reproduction: reproduction is the process by which organisms produce offspring. Different organisms have different methods of reproduction, including sexual and asexual reproduction.

Fertilization: fertilization is the process by which a male gamete (like a sperm) and a female gamete (like an egg) fuse together to form a zygote. This is the first stage of sexual reproduction.

Embryonic development: embryonic development is the process by which a fertilized egg becomes an embryo. This process includes cell division, differentiation and morphogenesis.

Pregnancy: pregnancy is when a woman carries a developing embryo or fetus in her uterus. The gestation period varies by species.

Parental care: parental care refers to the behaviors that parents exhibit when caring for their offspring, such as: b. Providing food, shelter and guidance.

Aging: aging is a process of aging and eventual death experienced by all organisms. Genetic and environmental factors can affect aging.

Life history strategies: life history strategies refer to a set of characteristics exhibited by an organism to optimize its survival and reproduction throughout its lifetime. Different organisms have different life history strategies depending on factors such as their environment, risk of predation, and reproductive opportunities.

Life chart: the life chart is a tool used by ecologists to analyze the survival and reproduction rates of populations over time. Life expectancy tables can provide insight into factors affecting population growth and survival.

Ecological succession: ecological succession is the process by which an ecosystem moves from one community to another over time. This process is driven by changes in abiotic and biotic factors such as climate, soil composition, and species interactions.

Rest: rest is a period of inactivity in the organism's life cycle, usually in response to unfavorable environmental conditions. Some plants and animals hibernate during periods of drought, extreme cold, or other stressful conditions to conserve energy and resources.

Hibernation: hibernation is a type of resting period that some animals go through during the winter months to conserve energy and survive harsh conditions. During hibernation, the animal's body temperature and metabolic rate drop, and the animal is able to sleep for longer periods of time.

Migration: migration is a behavior of some animals during their life cycle, moving from one place to another in search of resources such as food or breeding grounds. Migration can be short or long, involving several generations of animals.

Life cycle analysis: life cycle analysis is a method of assessing the environmental impact of a product or process throughout its life cycle, from raw material extraction to disposal. This method can provide information about the ecological advantages and disadvantages of different products and processes and form the basis for strategies for the sustainable use of resources.

Life cycle costing: life cycle costing is a method of estimating the costs associated with a product or process throughout its life cycle, including the costs of raw materials, production, use, maintenance and disposal. This method can provide insights into the financial merits and demerits of different products and processes, and help make decisions about resource use and waste management.

Population dynamics: population dynamics is the study of how populations of organisms change over time in response to factors such as birth rates, death rates, migration, and environmental conditions. Understanding population dynamics is important for managing and conserving natural resources and predicting the impact of human activities on ecosystems.

Understanding life cycles is important for understanding the biology and ecology of different organisms and for managing and protecting natural resources. Life history studies can provide insight into factors affecting population growth, survival, and reproductive success, which can inform strategies for conservation of threatened species and management of invasive species.

In general, life cycles are complex and dynamic processes involving a variety of biological, ecological and environmental factors. Understanding these factors and their interactions is essential for understanding the biology and ecology of different organisms, as well as for managing and protecting natural resources and reducing the impact of human activities on the environment.

Chapter No. 17

Human Body

Here are some concepts related to the human body:

anatomy: anatomy is the study of the structure and organization of the body and its organs. It includes the study of the skeletal system, muscular system, circulatory system, respiratory system, digestive system, nervous system and other systems.

Physiology: physiology is the study of how the body and its organs work, including the study of cell function, organ function and system function.

Cells: cells are the basic unit of life and the building blocks of tissues and organs. The human body contains trillions of cells, each with a specific function.

Tissues: tissues are groups of cells that work together to perform a specific function. Examples of tissues in the body include muscle tissue, neural tissue, and epithelial tissue.

Organs: organs are structures made up of different types of tissue that work together to perform a specific function. Examples of organs in the body are the heart, lungs, liver and brain.

Systems: systems are groups of organs that work together to perform a specific function. Examples of systems in the body include the circulatory system, respiratory system, digestive system, nervous system, and endocrine system.

Homeostasis: homeostasis is the body's ability to maintain a stable internal environment despite changes in the external environment. Factors such as body temperature, blood sugar and ph are regulated.

Immune system: the immune system is the body's defense against pathogens such as viruses and bacteria. It includes various cells, tissues and organs such as white blood cells, lymph nodes and spleen.

Reproductive system: the reproductive system is responsible for the production and transport of gametes (sperm and eggs) and for supporting fetal development during pregnancy.

Genetics: genetics is the study of genes and their role in inheritance and genetic variation. Includes tests for dna, chromosomes and genetic disorders.

Endocrine system: the endocrine system is responsible for the production and secretion of hormones that regulate various bodily functions, including growth and development, metabolism, and reproduction. The endocrine system includes glands such as the pituitary, thyroid, and adrenal glands.

Respiratory system: the respiratory system is responsible for the exchange of oxygen and carbon dioxide between the body and the environment. It includes organs such as the lungs, bronchi and trachea.

Digestive system: the digestive system is responsible for breaking down food and absorbing nutrients. This includes organs such as the mouth, esophagus, stomach, small intestine and large intestine.

Circulatory system: the circulatory system is responsible for moving blood, nutrients, oxygen, and waste products throughout the body. It includes the heart, blood vessels and blood.

Muscular system: the muscular system is responsible for generating movement and maintaining posture. It includes muscles like biceps, triceps and quadriceps.

Skeletal system: the skeletal system is responsible for supporting and protecting the body. It includes bones, cartilage and ligaments.

Nervous system: the nervous system is responsible for transmitting and processing information throughout the body. It includes the brain, spinal cord and nerves.

Sensory system: the sensory system is responsible for receiving and processing sensory information such as sight, hearing, taste, touch and smell.

Skin: the skin is the body's largest organ and is responsible for protecting the body from external damage, regulating body temperature and sensing touch.

Aging: aging is a process of biological changes that occur as a person ages. It includes changes in physical appearance, sensory abilities, cognitive function, and susceptibility to disease.

Overall, the human body is a complex and interconnected system, and each system and organ plays an important role in maintaining overall health and well-being. Understanding how each system works and how they are interconnected is critical to diagnosing and treating disease and promoting optimal health.

Chapter No. 18

Senses

The senses are the body's physiological ability to absorb information from the environment. The human body has five senses: sight, hearing, taste, smell and touch. Here are some concepts related to each sense:

sight: sight or sight is the ability of the eyes to perceive and interpret light. The eyes contain specialized cells called photoreceptors that are sensitive to light and help form visual images. The visual system also includes the optic nerve, which transmits visual information to the brain.

Hearing: hearing, or audition, is the ability of the ears to detect and interpret sound. The ears contain specialized cells called hair cells that are sensitive to sound vibrations and help to create auditory signals. The auditory system also includes the auditory nerve, which transmits auditory information to the brain.

Taste: taste, or gustation, is the ability of the tongue to detect and interpret flavors. The tongue contains specialized cells called taste buds that are sensitive to different types of molecules and help to create taste sensations. The taste system is also affected by other factors such as smell, texture and temperature.

Smell: smelling or smelling is the ability of the nose to perceive and interpret smells. The nose contains specialized cells called olfactory receptors that are sensitive to different types of molecules and help create the olfactory experience. The olfactory system is also closely related to the gustatory system and together they form the overall taste experience.

Touch: touch or somatosensing is the skin's ability to perceive and interpret different types of tactile stimuli such as pressure, temperature and pain. The skin contains specialized cells called sensory receptors that are sensitive to these stimuli and help create tactile sensations. The somatosensory system also includes nerves and pathways that transmit tactile information to the brain.

Each sense plays a key role in perceiving and interacting with the environment. They work together to create our overall perception of the world around us. Understanding how the different senses work and how they relate to each other is important for diagnosing and treating sensory disorders and developing technologies to improve sensory abilities.

Sensory integration: sensory integration is the process by which the brain receives and interprets information from the different senses and combines them into a cohesive perception of the environment. It is important for creating accurate and meaningful perceptions of the world around us.

Perception: perception is the process by which sensory information is organized, interpreted, and given meaning. It is influenced by factors such as attention, memory, and expectation, and can be influenced by individual differences such as culture and personality.

Sensory processing disorder: sensory processing disorder is a condition in which the brain has difficulty receiving and processing sensory information. It can cause a range of symptoms such as hypersensitivity, hypersensitivity, and problems with coordination and balance.

Synesthesia: synesthesia is a condition in which one sense activates another sense, resulting in a mixed perception of the environment. For example, a person with synesthesia may see colors when listening to music or taste aromas when seeing certain shapes.

Sensory substitution: sensory substitution is the use of one sense to replace or augment another impaired or absent sense. For example, a blind person can use auditory cues to navigate their surroundings using a technique called echolocation.

Understanding the complexity of the senses and how they interact with one another can provide insight into how we experience and perceive the world. Additionally, understanding sensory disorders and conditions can help medical professionals diagnose and treat people who may be struggling with sensory processing.

Chapter No. 19

Weather And Climate

Weather and climate are two related but different terms that describe the weather conditions in a specific area. Weather refers to the short-term atmospheric conditions at a specific location and time, while climate refers to the long-term average of those conditions.

Here are some key terms related to weather and climate:

temperature: temperature is a measure of how much any substance, including air, is heated or cooled. Temperature is affected by factors such as latitude, altitude and proximity to bodies of water.

Precipitation: precipitation is any form of moisture that falls from the atmosphere to the earth's surface, including rain, snow, sleet, and hail. Precipitation patterns can vary widely by region and season.

Humidity: humidity refers to the amount of water vapor in the air. Humidity can affect how hot or cold the air feels, as well as precipitation patterns.

Atmospheric pressure: atmospheric pressure is the weight of the atmosphere pressing on the earth's surface. Changes in barometric pressure can affect weather patterns and be a factor in thunderstorms.

Wind: wind refers to the movement of air from areas of high pressure to areas of low pressure. Wind patterns are influenced by a number of factors including temperature, pressure and the earth's rotation.

Climate zones: climate zones are large areas of the earth that exhibit similar weather and climate patterns. There are several climate zones including tropical, subtropical, temperate and polar.

Climate change: climate change refers to long-term changes in the earth's weather patterns, including rising global temperatures, changing precipitation patterns, and rising sea levels. Climate change is caused by a variety of human activities, including fossil fuel burning and deforestation.

Understanding weather and climate patterns is important in many applications, including agriculture, transportation, and disaster preparedness. Furthermore, understanding the impacts of climate change is crucial for the development of mitigation and adaptation strategies.

Weather forecasting: weather forecasting is the process of forecasting the weather conditions in a specific area at a specific time. It involves collecting and analyzing data from a variety of sources, including satellites, weather stations, and computer models.

El niño and la niña: el niño and la niña are weather conditions that occur in the pacific ocean and can have a significant impact on global weather patterns. El niño refers to the warming of the pacific ocean while la niña refers to the cooling of the pacific ocean.

Greenhouse effect: the greenhouse effect is a natural process that occurs when certain gases in the earth's atmosphere trap heat from the sun and keep the planet in a condition sufficient to support life. However, human activities have increased the concentration of these gases in the atmosphere, leading to increases in the greenhouse effect and global warming.

Global warming: global warming refers to the long-term increase in global temperatures that has occurred over the past century. This is mainly due to human activities such as burning fossil fuels, deforestation and industrial processes.

Climate adaptation: climate adaptation is the process of adapting to the impacts of climate change in order to minimize negative impacts and take advantage of potential opportunities. This can include strategies such as building levees to protect against sea level rise, developing drought-tolerant crops, and implementing heat warning systems in urban areas.

Extreme weather events: extreme weather events such as hurricanes, droughts and heat waves are becoming more frequent and more intense as a result of climate change. Understanding how climate change affects these events is critical to developing effective mitigation strategies.

Climate policy: climate policy refers to a set of laws, regulations and measures taken by governments and other organizations to counteract climate change. This can include measures such as carbon taxes, renewable energy commitments and emissions trading schemes.

Understanding weather and climate is a complex and ongoing process, so it is important that people and organizations are aware of the latest developments and trends. By taking action to combat and adapt to climate change, we can contribute to a more sustainable and resilient future.

Chapter No. 20

Water Cycle

The water cycle, also known as the water cycle, is a continuous process in which water circulates between the earth's surface and the atmosphere. The water cycle involves several key processes including evaporation, condensation, precipitation and infiltration.

Evaporation: evaporation is the process by which water changes from the liquid to the gaseous state. This happens when the sun's energy heats the surface of bodies of water, causing water molecules to gain energy and disperse in the air in the form of water vapor.

Condensation: condensation is the process by which water vapor cools and changes from gas to liquid. This happens when water vapor in the atmosphere meets cooler surfaces like clouds or the earth's surface and loses energy.

Precipitation: precipitation occurs when condensed water vapor falls to the earth's surface as rain, snow, sleet, or hail. This water can accumulate in bodies of water such as rivers, lakes and oceans, or seep into the ground as groundwater.

Infiltration: infiltration is the process by which water is absorbed into the soil and becomes groundwater. This water can be stored in underground aquifers and eventually rise to the surface through springs or other forms of aquifer drainage.

Transpiration: transpiration is the process by which plants absorb water from the soil and release it into the atmosphere through the tiny pores of the leaves. This water then becomes part of the water cycle and may eventually fall to the surface of the earth as precipitation.

The hydrological cycle is a key process for sustaining life on earth as it ensures the constant availability of water for plant, animal and human communities. However, human activities such as deforestation, land-use change and pollution can have significant impacts on the water cycle, leading to water scarcity, droughts and other environmental problems. Therefore, it is important that individuals and organizations take action to protect and conserve water resources and promote sustainable practices that support the health of the water cycle and the planet as a whole.

In addition to the key processes involved in the water cycle, there are also several factors that can affect the movement and distribution of water on earth:

topography: the topography of the earth's surface can affect the movement of water by affecting the direction and speed of water flow. For example, water naturally flows downward and tends to pool in low-lying areas such as valleys and pools.

Climate: the climate plays an important role in the water cycle and affects the rate of evaporation and precipitation. Areas of high temperature and high humidity tend to have higher rates of evaporation, while areas of lower temperature and higher elevation tend to have higher rates of precipitation.

Vegetation: vegetation can affect the water cycle by affecting the rate of transpiration and infiltration. Plants draw water from the soil through their roots and release it into the atmosphere through their leaves. This process can help maintain moisture levels in the atmosphere and prevent droughts.

Human activities: human activities such as agriculture, industry and urbanization can have significant impacts on the water cycle. For example, deforestation can lead to soil erosion and reduced infiltration rates, while urbanization can lead to increased runoff and reduced groundwater recharge.

Climate change: climate change is expected to have significant impacts on the hydrological cycle, with projections indicating changes in precipitation patterns, evaporation rates and water availability in many parts of the world. These changes can have significant impacts on ecosystems, water resources and human communities.

Understanding the factors that affect the water cycle is important to managing water resources and protecting the health of our planet. By taking action to conserve water, promote sustainable practices and mitigate the effects of climate change, individuals and organizations can help maintain a healthy and thriving water cycle for future generations.

Chapter No. 21

Earth And Space

Earth and space sciences cover a wide range of topics, from the composition and structure of the earth and its atmosphere to the formation and evolution of the universe. Some key areas of earth and space science research are:

geology: geology is the study of the physical structure, composition and history of the earth. Geologists study topics such as plate tectonics, the formation of rocks and minerals, and processes that shape the earth's surface, such as erosion and weathering.

Meteorology: meteorology is the study of the earth's atmosphere and weather conditions. Meteorologists use a variety of tools and techniques to observe, model, and forecast weather conditions, including temperature, precipitation, and wind.

Astronomy: astronomy is the study of the universe beyond the earth. Astronomers study such subjects as the formation and evolution of stars and galaxies, the properties of planets and other celestial bodies, and the origin of the universe itself.

Oceanography: oceanography is the study of the earth's oceans, including their physical properties, chemical composition, and ecosystems. Oceanographers study topics such as ocean currents, marine ecosystems, and the impact of human activities on ocean health.

Environmental science: environmental science is the study of the interactions between human communities and the natural environment. Environmental scientists deal with issues such as climate change, pollution and resource management and work on solutions to the ecological challenges of our planet.

Understanding earth and space science is key to addressing the many pressing challenges facing our planet, from climate change and resource depletion to natural disasters and environmental degradation. By studying the complex systems and processes that shape our planet and the universe beyond, scientists can help develop strategies to protect the health and well-being of our planet and all of its inhabitants.

Earthquakes and volcanoes: earthquakes and volcanoes are natural phenomena closely linked to the movement of tectonic plates, large pieces of the earth's crust that fit together like a jigsaw puzzle. Earthquakes are caused by the sudden release of energy stored in the earth's crust, while volcanoes are caused by the eruption of molten rock and ash from the earth's mantle.

Climate change: climate change refers to long-term changes in global temperature, precipitation patterns, and other weather variables that are largely caused by human activities such as fossil fuel burning and deforestation. Climate change has significant impacts on the earth's ecosystems, water resources and human society and is one of the greatest ecological challenges facing our planet.

Planetary science: planetary science is the study of the planets and other celestial bodies in our solar system and beyond. Planetary scientists study topics such as planet formation and evolution, the search for extraterrestrial life, and the potential for human exploration and colonization of other planets.

Earth's magnetic field: the earth's magnetic field is created by the movement of molten iron in the earth's core and plays a key role in protecting the planet from the harmful effects of solar radiation. The magnetic field also aids in animal migration and is used by humans for navigation and communication.

Space exploration: space exploration refers to the scientific study of outer space, including the use of spacecraft and other technology to study planets, moons, asteroids, and planets, other celestial bodies in our solar system and beyond. Space exploration has led to many important discoveries and advances in our understanding of the universe, inspiring generations of scientists and explorers.

Overall, earth and space science is a fascinating and dynamic field that encompasses many different areas of research, from how our planet works to the mysteries of the universe behind it.by continuing to study and understand these issues, scientists can help develop solutions to the environmental and social challenges facing our planet and inspire new generations to discover the wonders of the universe.

Chapter No. 22

Sun, Moon, And Stars

Here are some additional topics and concepts related to the sun, moon, and stars:

solar system: the solar system is the collection of planets, moons, asteroids, comets, and other celestial bodies orbiting the sun. The eight planets in our solar system are mercury, venus, earth, mars, jupiter, saturn, uranus and neptune. Each planet has unique characteristics such as size, composition and atmospheric conditions, making it an intriguing subject for scientists to study.

Phases of the moon: the moon goes through a cycle of phases caused by its orbit around the earth and the changing angle of sunlight illuminating its surface. The phases include new moon, waxing crescent, first quarter, waxing bulge, full moon, waning bulge, third quarter and waning crescent. Understanding the phases of the moon is important to various cultural practices, such as b. Determining the beginning of a new month in some calendars.

Solar and lunar eclipses: solar and lunar eclipses occur when the sun, moon and earth are aligned in a certain way. A solar eclipse occurs when the moon passes between the sun and the earth, blocking sunlight and casting a shadow on the earth's surface. A lunar eclipse occurs when the earth passes between the sun and moon, casting a shadow on the lunar surface. Both types of solar eclipses are rare events that have fascinated people for centuries.

Constellations: constellations are groups of stars that appear to form patterns or shapes in the night sky. There are 88 officially recognized constellations, each with its own history and cultural significance. Constellations have been used for navigation, storytelling, and astrological purposes throughout human history.

Stellar evolution: stars are born from clouds of gas and dust in space and undergo a complex evolutionary process over millions or billions of years. A star's life cycle depends on its mass, with more massive stars having shorter lifetimes and more dramatic final stages. Understanding the life cycle of stars is important to understanding the chemical and physical processes that shape our universe.

Solar wind: the solar wind is a stream of charged particles (mainly protons and electrons) that emanates from the sun and permeates the entire solar system. The solar wind can interact with the earth's magnetic field and cause phenomena such as the aurora borealis (also called the northern lights and aurora borealis).

Star classification: astronomers classify stars based on their temperature, size, brightness, and other properties. The most commonly used classification system is the hertzsprung-russell (hr) diagram, which represents stars. Brightness as a function of their temperature. This allows astronomers to understand how stars evolve over time and how they contribute to the overall structure and dynamics of the universe.

Nebulae: nebulae are clouds of gas and dust in space from which new stars can form. Some of the most famous nebulae are the orion nebula and the eagle nebula. Nebulae can come in a variety of shapes, sizes, and colors, and are important objects of study for astronomers to understand the formation and evolution of stars and galaxies.

Black holes: black holes are extremely dense objects that are formed when massive stars collapse into one another. They have gravitational fields so strong that nothing, not even light, can escape once it crosses the "event horizon"." black holes are difficult to observe directly, but scientists can study their effects on surrounding matter to learn more about their properties and behavior.

Cosmology: cosmology is the study of the origin, structure and development of the universe as a whole. Cosmologists use observations of galaxies, the cosmic microwave background, and other phenomena to develop theories about the early universe, dark matter, dark energy, and other aspects of the cosmos. Cosmology is an exciting and fast-growing field of astronomy that continues to push the boundaries of our understanding of the universe.

The sun, moon, and stars taken together are fascinating objects that have excited human imagination and scientific curiosity for thousands of years. By continuing to study and study these celestial bodies, scientists can learn more about the origin and evolution of our solar system and the entire universe.

Chapter No. 23

Rocks And Minerals

Here are some additional topics and concepts related to rocks and minerals:

properties of minerals: minerals are naturally occurring substances that have chemical compositions and physical properties. Some of the most important physical properties of minerals are color, luster, hardness, cleavage, and fracture. These properties can help geologists identify and classify different types of minerals.

Rock types: rocks are made up of minerals and can be divided into three main categories based on how they were formed: igneous, sedimentary and metamorphic. Igneous rocks are formed from solidified magma or lava, sedimentary rocks are formed from the accumulation of sediment and organic material, and metamorphic rocks are formed from the transformation of existing rocks by heat and pressure.

Formation of minerals: minerals can be formed by a variety of processes, including crystallization from magma or solution, precipitation from thermal springs or mineral-rich groundwater, and biomineralization (where organisms form minerals). Understanding the processes that produce different types of minerals can provide insight into the geological history of an area.

Use of mincrals: minerals are important raw materials that are used in a variety of ways in industry, technology and everyday life. For example, quartz is used in electronics and glass, copper in electrical and plumbing wiring, and gypsum in building materials.

Rock cycle: the rock cycle is an ongoing process in which rocks change from one type to another over geologic time. This process involves the formation, erosion and metamorphism of rock and is driven by a combination of internal and external forces such as heat, pressure, water, wind and gravity. Understanding the rock cycle can provide insight into the history and evolution of the earth's crust.

Mineral resources: mineral resources are minerals extracted from the earth and used commercially. Typical mineral resources are metals such as copper, gold and iron as well as non-metallic commodities such as coal, oil and natural gas. The extraction and use of mineral resources can have significant environmental and social impacts, so it is important to manage these resources sustainably to ensure their availability for future generations.

Mineral deposits: mineral deposits are groups of minerals that are economically viable to extract. These deposits can be found in a variety of geological environments such as veins, sedimentary layers, and igneous intrusions.

Rock formation: rocks can form in a variety of ways, including volcanic eruptions, sedimentary deposition, and metamorphism. Understanding the processes that create different rock types can provide insight into the geological history of an area.

Mineralogy: mineralogy is the study of minerals, including their chemical composition, physical properties and processes of formation. Mineralogists use a variety of techniques to study minerals such as b. X-ray diffraction and spectroscopy.

Geology: geology is the study of the structure, processes and history of the earth. Geologists use a variety of tools and methods to study the earth, including remote sensing, seismic imaging, and field observations. The study of geology is important for understanding natural hazards such as earthquakes and volcanic eruptions and for the sustainable management of natural resources.

Crystallography: crystallography is the study of the structure of crystals, including their shape, symmetry and arrangement of atoms. Understanding crystallography is important for mineralogists and materials scientists because it can provide insight into the physical properties and behavior of minerals and other crystalline materials.

Ore deposits: ore deposits are ore deposits that contain precious metals or economically mineable ores. Some common types of mineral deposits include hydrothermal deposits, porphyry deposits, and alluvial deposits. The discovery and exploitation of mineral deposits has played an important role in human history and economic development.

Petrology: petrology is the study of rocks, including their composition, structure and formation processes. Petrologists use a variety of techniques to study rocks, such as: b. Thin section analysis and petrographic microscopy.

Volcanology: volcanology is the study of volcanoes and volcanic processes, including eruptions, magma chambers, and lava flows. Volcanologists use a variety of tools and methods to study volcanoes, such as b. Seismology, gas monitoring and field observations. Understanding volcanology is important to predict volcanic hazards and manage volcanic risk.

Geochronology: geochronology is the study of the age of the earth and rocks using techniques such as radiometric dating and stratigraphy. Geochronology can provide information about the geological history of the earth and the evolution of life on our planet.

Mineral exploration: mineral exploration is the process of finding new mineral deposits. This process involves a variety of techniques such as geological mapping, geophysical surveys and drilling. Mineral exploration is important to identify new mineral resources and ensure a sustainable supply of minerals for commercial and industrial use.

Mineralogy in space: the study of minerals is not limited to earth as minerals can also be found in space. Meteorites and other extraterrestrial materials contain a variety of minerals that can provide insight into the formation and evolution of the solar system.

Types of minerals: there are over 5,000 known minerals on earth, each with their own unique properties and characteristics. Studying the diversity of minerals can provide insight into the chemical and physical processes that shape the earth's crust.

Mineralogy and society: the study of minerals has played an important role in human history and development, from the ancient use of flint to make tools to the modern use of minerals in technology and industry. Understanding the social and economic impacts of mineral use is important for the sustainable management of mineral resources and ensuring equitable access to these resources.

Chapter No. 24

Careers In Science

Here are some scientific professions:

biologist: biologists study living organisms and their interactions with their environment. You can work in fields such as ecology, genetics, microbiology or biochemistry.

Chemist: chemists study the composition, structure, and properties of matter, including their chemical reactions and interactions. You can work in fields such as pharmacy, materials science, or environmental science.

Physicists: physicists study the fundamental laws of nature, including how matter and energy behave at the atomic and subatomic levels. You can work in fields such as astrophysics, particle physics or condensed matter physics.

Geologist: geologists study the structure, composition, and history of the earth. You may work in fields such as mineral exploration, environmental science, or petroleum geology.

Environmental scientists: environmental scientists study the natural environment and its interaction with human activities. You can work in areas such as conservation, sustainability or environmental protection.

Medical scientist: medical scientists study the causes, diagnosis and treatment of diseases. You can work in fields such as pharmacology, epidemiology or biotechnology.

Data scientist: data scientists use statistical and computational methods to analyze and interpret large amounts of data. You can work in areas such as finance, health or social media.

Astronomer: astronomers study the universe, including its origin, composition and evolution. You can work in fields such as astrophysics, cosmology or planetary science.

Marine scientists: marine scientists study the oceans and their inhabitants, including marine life, ocean currents, and marine ecosystems. You can work in fields such as oceanography, marine biology or fisheries science.

Science teachers: science teachers educate students at different levels, from elementary school through college. You can work in schools, museums or science centers.

Science writer: science writers communicate scientific information to a variety of audiences, including the general public, policy makers, and other scientists. You can work for scientific journals, news organizations or science communication agencies.

Patent attorney: patent attorneys specialize in intellectual property law, including patents for scientific inventions. They may work in law firms, corporations, or government agencies.

Forensic technician: forensic scientists use scientific methods to analyze evidence in criminal investigations, including dna, fingerprints, and other types of physical evidence. You can work for law enforcement agencies, government labs, or private companies.

Biomedical engineer: biomedical engineers apply engineering principles to develop new medical devices and technologies. You may work in fields such as prosthetics, medical imaging, or drug delivery.

Science policy advisors: science policy advisors work with policy makers to develop and implement science-based policies. You can work for government agencies, non-profit organizations, or think tanks.

Chapter No. 25

Science Experiments At Home

Science experiments at home are a great way for kids to learn about scientific methods and explore different scientific concepts in a fun and engaging way. Here are some science experiment ideas you can do at home:

baking soda and vinegar volcano: mix baking soda and vinegar to create a chemical reaction that produces carbon dioxide. This causes a volcanic eruption.

Sink or float: fill a container with water and collect various household items to see if they sink or float. This experience helps children learn buoyancy.

Ice melting contest: fill two containers with water and place an ice cube in each. Place one container in the freezer and leave the other at room temperature. The ice cube in the freezer melts longer, so children learn about heat transfer.

Home slime: mix glue, water and borax to make slime. This experiment teaches children chemical reactions and polymerization.

Static balloon: rub a balloon on a wool sweater to generate static electricity. Hold the balloon near different materials to see which ones are attracted. This experiment helps children discover electricity and magnetism.

Egg in a bottle: place a pcclcd, hard-boiled egg in a small-mouthed bottle. Light a piece of paper and put it in the bottle. As the paper burns, the egg is sucked into the bottle. This experiment shows atmospheric pressure.

Rainbow milk: place drops of food coloring in a deep plate with the milk. Then put a drop of washing-up liquid in the middle of the plate. Dish soap swirls and mixes colors to create a colorful display. This experiment teaches children about surface tension.

Mentos & sparkling geyser: drop some mentos candies into a soda bottle to create a geyser. This experiment teaches children about chemical reactions and carbon dioxide.

Home lava lamp: fill a clear vessel with oil and water. Add a few drops of food coloring and an alka-seltzer tablet. As the tablet dissolves it creates bubbles that rise and fall creating a lava lamp effect. This experiment teaches children about density and chemical reactions.

Invisible ink: write a message with lemon juice on a piece of white paper. Allow the paper to dry, then place it over a heat source (such as a lightbulb or iron). The heat turns the lemon juice brown, revealing a hidden message. This experiment teaches children acid-base reactions.

Make your own electromagnet: wrap a copper wire around a nail and connect the ends of the wire to a battery. The nail becomes magnetized and can pick up paperclips and other small metal objects. This experiment teaches children about electricity and magnetism.

Homemade stethoscope: use a cardboard tube, funnel, and a balloon to create a homemade stethoscope. Place the funnel on the chest and listen through the cardboard tube to hear the heart beating. This experiment teaches children about the human body and sound waves.

Density tower: fill the transparent container with liquids of different densities such as corn syrup, water and oil. The liquids stack on top of each other to create a colorful and interesting display. This experiment teaches children about density and buoyancy.

Solar cooker: create a solar cooker by covering a cardboard box with aluminum foil and adding a clear plastic lid. Put the food in the oven and leave it in the sun for a few hours. Solar energy cooks the food. This experiment teaches children about renewable energy and heat transfer.

Acid-base reaction: mix baking soda and vinegar to create a chemical reaction.

Add food coloring to the mixture to make it more visually appealing. This experiment teaches children chemical reactions and acid-base reactions.

Edible dna: use candies to create a dna model. Use different colors of liquor ice or gum to represent different dna components. This experiment teaches kids about genetics and dna.

Egg drop challenge: encourage kids to build a protective container for eggs so they can be dropped from a great height without breaking. Use a variety of materials to craft the container such as: b. Cardboard, bubble wrap and cotton swabs. This experience teaches children technique and strength.

Crystal growing: dissolve sugar or salt in hot water and allow to cool. Put a piece of string or a wooden stick in the solution and let it sit for a few days. Crystals grow on a string or stick. This experiment teaches children chemical reactions and crystal formation.

Dry ice blowing: put dry ice in a container, add water and soap. When dry ice sublimates, it creates a cloud of carbon dioxide that fills the container and creates bubbles. This experiment teaches children about sublimation and carbon dioxide.

Balloon rocket: tie the string to the balloon and attach to the straw. Inflate the balloon and release it to make it fly across the room. This experience teaches children about strength and movement.

These are just a few ideas for science experiments at home. Encourage children to develop their own ideas and experiment with different materials and designs.

Remember to always supervise children when doing science experiments and to take all precautions.